John Dalton, William Hyde Wollaston, Thomas Thomson

Foundations of the Atomic Theory

John Dalton, William Hyde Wollaston, Thomas Thomson

Foundations of the Atomic Theory

ISBN/EAN: 9783337399030

Printed in Europe, USA, Canada, Australia, Japan

Cover: Foto ©berggeist007 / pixelio.de

More available books at **www.hansebooks.com**

FOUNDATIONS

OF THE

ATOMIC THEORY

COMPRISING

PAPERS AND EXTRACTS

BY

JOHN DALTON
WILLIAM HYDE WOLLASTON, M.D.
AND
THOMAS THOMSON, M.D.
(1802-1808)

Edinburgh
THE ALEMBIC CLUB

Chicago
THE UNIVERSITY OF CHICAGO PRESS
1911

PREFACE.

THIS little book contains reprints of original memoirs and extracts from text-books, embracing the earliest publications by their respective authors bearing upon the foundation of the Atomic Theory.

The view is pretty generally held by chemists that it was in the endeavour to explain numerous examples which were known to him, of that general regularity which is now commonly called the Law of Multiple Proportions, that Dalton was led to entertain the ideas which he held regarding the constitution of compound bodies. There has therefore been included, along with later publications, the paper by Dalton in which there is described probably the first example of this regularity with which he became acquainted.

The first part of Dalton's "New System of Chemical Philosophy," containing his own account of his views, did not appear until 1808, although these views had been embodied in courses of lectures which Dalton had delivered some years previously. The earliest printed account of his views is that given by Dr Thomas Thomson in Volume 3 of the Third Edition of his "System of Chemistry," published in 1807. This account is reproduced in the following pages.

A paper by Wollaston on Super-acid and Sub-acid Salts, giving various examples illustrative of the Law of Multiple Proportions, is also included.

L. D.

EXPERIMENTAL ENQUIRY INTO THE PROPORTION OF THE SEVERAL GASES OR ELASTIC FLUIDS, CONSTITUTING THE ATMOSPHERE. By JOHN DALTON.*

Read Nov. 12, 1802.

———❈———

IN a former paper which I submitted to this Society, "On the constitution of mixed gases," I adopted such proportions of the simple elastic fluids to constitute the atmosphere as were then current, not intending to warrant the accuracy of them all, as stated in the said paper; my principal object in that essay was, to point out the *manner* in which mixed elastic fluids exist together, and to insist upon what I think a very important and fundamental position in the doctrine of such fluids :— namely, that the elastic or repulsive power of each particle is confined to those of its own kind; and consequently the force of such fluid, retained in a given vessel, or gravitating, is the same in a separate as in a mixed state, depending upon its proper density and temperature. This principle accords with all experience, and I have no doubt will soon be perceived and acknowledged by chemists and philosophers in general; and its application will elucidate a variety of facts, which are otherwise involved in obscurity.

* From the Memoirs of the Literary and Philosophical Society of Manchester, Second Series, Volume I., 1805, pp. 244-258. In this paper there is announced the first example of the law of multiple proportions.

The objects of the present essay are,

1. To determine the weight of each simple atmosphere, abstractedly; or, in other words, what part of the weight of the whole compound atmosphere is due to azote; what to oxygen, &c. &c.

2. To determine the relative weights of the different gases in a given volume of atmospheric air, such as it is at the earth's surface.

3. To investigate the proportions of the gases to each other, such as they ought to be found at different elevations above the earth's surface.

To those who consider the atmosphere as a chemical compound, these *three* objects are but *one :* others, who adopt my hypothesis, will see they are essentially distinct.— With respect to the first: It is obvious, that, on my hypothesis, the density and elastic force of each gas at the earth's surface, are the effects of the weight of the atmosphere of that gas solely, the different atmospheres not gravitating one upon another. Whence the first object will be obtained by ascertaining what share of elastic force is due to each gas in a given volume of the compound atmosphere; or, which amounts to the same thing, by finding how much the given volume is diminished under a constant pressure, by the abstraction of each of its ingredients singly. Thus, if it should appear that by extracting the oxygenous gas from any mass of atmospheric air, the whole was diminished $\frac{1}{5}$ in bulk, still being subject to a pressure of 30 inches of mercury; then it ought to be inferred that the oxygenous atmosphere presses the earth with a force of 6 inches of mercury, &c.

In order to ascertain the second point, it will be further necessary to obtain the specific gravity of each gas; that is, the relative weights of a given volume of each in a pure state, subject to the same pressure and temperature. For, the weight of each gas in any given portion of atmospheric air, must be in the compound ratio of its force and specific gravity.

With respect to the third object, it may be observed, that those gases which are specifically the heaviest must decrease in density the quickest in ascending. If the earth's atmosphere had been a homogeneous elastic fluid of the same weight it is, but ten times the specific gravity, it might easily be demonstrated that no sensible portion of it could have arisen to the summits of the highest mountains. On the other hand, an atmosphere of hydrogenous gas, of the same weight, would support a column of mercury nearly 29 inches on the summit of Mount Blanc.

The several gases constantly found in every portion of atmospheric air, and in such quantities as are capable of being appreciated, are azotic, oxygenous, aqueous vapour, and carbonic acid. It is probable that hydrogenous gas also is constantly present; but in so small proportion as not to be detected by any test we are acquainted with; it must therefore be confounded in the large mass of azotic gas.

1. *Of the Weight of the Oxygenous and Azotic Atmospheres.*

Various processes have been used to determine the quantity of oxygenous gas.

1. The mixture of nitrous gas and air over water.

2. Exposing the air to liquid sulphuret of potash or lime, with or without agitation.

3. Exploding hydrogen gas and air by electricity.

4. Exposing the air to a solution of green sulphat or muriat of iron in water, strongly impregnated with nitrous gas.

5. Burning phosphorus in the air.

In all these cases the oxygen enters into combination and loses its elasticity; and if the several processes be conducted skilfully, the results are precisely the same from all. In all parts of the earth and at every season of the year, the bulk of any given quantity of atmospheric air appears to be reduced nearly 21 per cent. by abstracting its oxygen. This fact, indeed, has not been generally admitted till lately; some chemists having found, as they apprehended, a great difference in the quantity of oxygen in the air at different times and places; on some occasions 20 per cent. and on others 30, and more of oxygen are said to have been found. This I have no doubt was owing to their not understanding the nature of the operation and of the circumstances influencing it. Indeed it is difficult to see, on any hypothesis, how a disproportion of these two elements should ever subsist in the atmosphere.

As the first of the processes above-mentioned has been much discredited by late authors, and as it appears from my experience to be not only the most elegant and expeditious of all the methods hitherto used, but also as correct as any of them, when properly conducted, I shall, on this occasion, animadvert upon it.

1. Nitrous gas may be obtained pure by nitric acid diluted with an equal bulk of water poured upon copper or mercury; little or no artificial heat should be applied.— The last product of gas this way obtained, does not contain any sensible portion of azotic gas; at least it may easily be got with less than 2 or 3 per cent. of that gas : It is probably nearly free from nitrous oxide also, when thus obtained.

2. If 100 measures of common air be put to 36 of pure nitrous gas in a tube 3·10th of an inch wide and 5 inches long, after a few minutes the whole will be reduced

to 79 or 80 measures, and exhibit no signs of either oxygenous or nitrous gas.

3. If 100 measures of common air be admitted to 72 of nitrous gas in a wide vessel over water, such as to form a thin stratum of air, and an immediate momentary agitation be used, there will, as before, be found 79 or 80 measures of pure azotic gas for a residuum.

4. If, in the last experiment, *less* than 72 measures of nitrous gas be used, there will be a residuum containing oxygenous gas; if *more*, then some residuary nitrous gas will be found.

These facts clearly point out the theory of the process: the elements of oxygen may combine with a certain portion of nitrous gas, or with twice that portion, but with no intermediate quantity.   In the former case *nitric* acid is the result; in the latter *nitrous* acid : but as both these may be formed at the same time, one part of the oxygen going to *one* of nitrous gas, and another to *two*, the quantity of nitrous gas absorbed should be variable; from 36 to 72 per cent. for common air.   This is the principal cause of that diversity which has so much appeared in the results of chemists on this subject.   In fact, all the gradation in quantity of nitrous gas from 36 to 72 may actually be observed with atmospheric air of the same purity; the wider the tube or vessel the mixture is made in, the quicker the combination is effected, and the more exposed to water, the greater is the quantity of *nitrous* acid and the less of *nitric* that is formed.

To use nitrous gas for the purpose of eudiometry therefore, we must attempt to form *nitric acid* or *nitrous* wholly, and without a mixture of the other.   Of these the former appears from my experiments to be most easily and most accurately effected.   In order to this a narrow tube is necessary; one that is just wide enough to let air pass water without requiring the tube to be agitated, is best. Let little more nitrous gas than is sufficient to form nitric acid be admitted to the oxygenous gas; let no agitation be used; and as soon as the diminution appears to be over for a moment let the residuary gas be transferred to another tube, and it will remain without any further diminution of consequence. Then $\frac{7}{19}$ of the loss will be due to oxygen.—The transferring is necessary to prevent the nitric acid formed and combined with the water, from absorbing the remainder of the nitrous gas to form nitrous acid.

Sulphuret of lime is a good test of the proportion of oxygen in a given mixture, provided the liquid be not more than 20 or 30 per cent. for the gas (atmospheric air); if the liquid exceed this, there is a portion of azotic gas imbibed somewhat uncertain in quantity.

Volta's eudiometer is very accurate as well as elegant and expeditious: according to Monge, 100 oxygen require 196 measures of hydrogen; according to Davy 192; but from the most attentive observations of my own, 185 are sufficient. In atmospheric air I always find 60 per cent. diminution when fired with an excess of hydrogen; that is, 100 common air with 60 hydrogen, become 100 after the explosion, and no oxygen is found in the residuum; here 21 oxygen take 39 hydrogen.

2. *Of the Weight of the Aqueous Vapour Atmosphere.*

I have, in a former essay, (Manchester Mem. vol. 5. p. 2, page 559.) given a table of the force of vapour in *vacuo* for every degree of temperature, determined by experiment; and in the sequel of the essay, have shown that the force of vapour in the atmosphere is the very same as in *vacuo*, when they are both at their utmost for any given temperature. To find the force of aqueous vapour in the atmosphere, therefore, we have nothing

.nore to do than to find that degree of cold at which it begins to be condensed, and opposite to it in the table above mentioned, will be found the force of vapour. From the various facts mentioned in the essay it is obvious, that vapour contracts no chemical union with any of the gases in the atmosphere ; this fact has since been enforced in the Annales de Chimie, vol. xlii. by Clement and Desorme.

M. De Saussure found by an excellent experiment, that dry air of 64° will admit so much vapour as to increase its elasticity, $\frac{1}{54}$.—This I have repeated nearly in his manner, and found a similar result. But the table he has given us of aqueous vapour at other temperatures is very far wrong, especially at temperatures distant from 64°.—The numbers were not the result of direct experiment, like the one above.—If we could obtain the temperatures of all parts of the earth's surface, for any given time, a mean of them would probably be 57° or 58°. Now if we may suppose the force of vapour equivalent to that of 55°, at a medium, it will, from the table, be equal to .443 of mercury ; or, nearly $\frac{1}{70}$ of the whole atmosphere. This it will be perceived is calculated to be the weight of vapour in the whole atmosphere of the earth. If that incumbent over any place at any time be required, it may be found as directed above.

3. *Of the Weight of the Carbonic Acid Atmosphere.*

From some observations of Humboldt, I was led to expect about $\frac{1}{100}$ part of the weight of the atmosphere to be carbonic acid gas : but I soon found that the proportion was immensely overrated. From repeated experiments, all nearly agreeing in their results, and made at different seasons of the year, I have found, that if a glass vessel filled with 102,400 grains of rain water be emptied in the open air, and 125 grains of strong lime water be poured in, and the mouth then closed; by sufficient time and agitation, the whole of the lime water is just saturated by the acid gas it finds in that volume of air. But 125 grains of the lime water used require 70 grain measures of carbonic acid gas to saturate it; therefore, the 102,400 grain measures of common air contain 70 of carbonic acid; or $\frac{1}{1460}$ of the whole.—The weight of the carbonic acid atmosphere then is to that of the whole compound as 1 : 1460; but the weight of carbonic acid gas in a given portion of air at the earth's surface, is nearly $\frac{1}{1000}$ of the whole; because the specific gravity of the gas is $1\frac{1}{2}$ that of common air. I have since found that the air in an assembly, in which two hundred people had breathed for two hours, with the windows and doors shut, contained little more than 1 per cent. of carbonic acid gas.

Having now determined the force with which each atmosphere presses on the earth's surface, or in other words, its weight; it remains next to enquire into their specific gravities.

These may be seen in the following Table.

Atmospheric air,	-	-	-	1.000
Azotic gas,	-	-	-	.966
Oxygenous gas,	-	-	-	1.127
Carbonic acid gas,	-	-	-	1.500
Aqueous vapour,	-	-	-	.700
Hydrogenous gas,	-	-	-	.077*

Kirwan and Lavoisier are my authorities for these numbers; except oxygenous gas and aqueous vapour.

---

* The specific gravity of hydrogen must be rated too low: if 100 oxygen require 185 hydrogen by measure, according to this 89 oxygen would require only 11 hydrogen to form water; whereas 85 require 15. Hydrogen ought to be found about $\frac{1}{10}$ part of the weight of common air.

For the former I am indebted to Mr Davy's Chemical Researches; his number is something greater than theirs: I prefer it, because, being determined with at least equal attention to accuracy with the others, it has this further claim for credit, that 21 parts of gas of this specific gravity, mixed with 79 parts of azotic gas, make a compound of exactly the same specific gravity as the atmosphere, as they evidently ought to do, setting aside the unfounded notion of their forming a *chemical* compound. The specific gravity of aqueous vapour I have determined myself both by analytic and synthetic methods, after the manner of De Saussure; that is, by abstracting aqueous vapour of a known force from a given quantity of air, and weighing the water obtained—and admitting a given weight of water to dry air and comparing the loss with the increased elasticity. De Saussure makes the specific gravity to be ,71 or ,75; but he used caustic alkali as the absorbent, which would extract the carbonic acid as well as the aqueous vapour from the air. From the experiments of Pictet and Watt, I deduce the specific gravity of aqueous vapour to be ,61 and ,67 respectively. Upon the whole, therefore, it is probable that ,7 is very nearly accurate.

We have now sufficient data to form tables answering to the two first objects of our enquiry.

I. *Table of the Weights of the different Gases constituting the Atmosphere.*

	Inch of Mercury.
Azotic gas - - - -	23.36
Oxygenous gas - - - -	6.18
Aqueous vapour - - -	.44
Carbonic acid gas - - -	.02
	30.00

II. *Table of the proportional Weights of the different Gases in a given volume of Atmospheric Air, taken at the Surface of the Earth*

				Per Cent.
Azotic gas	-	-	-	75.55
Oxygenous gas	-	-	-	23.32
Aqueous vapour	-	-	-	1.03*
Carbonic acid gas		-	-	.10

100.00

III. *On the Proportion of Gases at different Elevations.*

M. Berthollet seems to think that the lower strata of the atmosphere ought to contain more oxygen than the upper, because of the greater specific gravity of oxygenous gas, and the slight affinity of the two gases for each other. (See Annal. de Chimie, Tom. 34. page 85.) As I am unable to conceive even the possibility of two gases being held together by affinity, unless their particles unite so as to form *one* centre of repulsion out of two or more (in which case they become *one* gas) I cannot see why rarefaction should either decrease or increase this supposed affinity. I have little doubt, however, as to the fact of oxygenous gas observing a diminishing ratio in ascending; for, the atmospheres being independent on each other, their densities at different heights must be regulated by their specific gravities.—Hence, if we take the azotic atmosphere as a standard, the oxygenous and the carbonic acid will observe a decreasing ratio to it in ascending, and the aqueous vapour an increasing one. The specific gravity of oxygenous and azotic gases being as 7 to 6 nearly, their diminution in density will be the same at heights reciprocally as their specific gravities.

---

* The proportion of aqueous vapour must be understood to be variable for any one place : the others are permanent or nearly so.

Hence it would be found, that at the height of Mount Blanc (nearly three English miles) the ratio of oxygenous gas to azotic in a given volume of air, would be nearly as 20 to 80 ;—consequently it follows that at any ordinary heights the difference in the proportions will be scarcely if at all perceptible.*

ON THE ABSORPTION OF GASES BY WATER AND OTHER LIQUIDS. By JOHN DALTON.†

Read Oct. 21, 1803.

1. IF a quantity of pure water be boiled rapidly for a short time in a vessel with a narrow aperture, or if it be subjected to the air-pump, the air exhausted from the receiver containing the water, and then be briskly agitated for some time, very nearly the whole of any gas the water may contain, will be extricated from it.

2. If a quantity of water thus freed from air be agitated in any kind of gas, not chemically uniting with water, it will absorb its bulk of the gas, or otherwise a part of it equal to some one of the following fractions, namely, $\frac{1}{8}$, $\frac{1}{27}$, $\frac{1}{64}$, $\frac{1}{125}$, &c. these being the cubes of the reciprocals of the natural numbers 1, 2, 3, &c. or $\frac{1}{1}$ $\frac{1}{2^3}$, $\frac{1}{3^3}$, $\frac{1}{4^3}$, &c. the same gas always being absorbed in the same propor-

---

* Air brought from the summit of Helvelyn, in Cumberland (1100 yards above the sea—Barometer being 26.60) in July 1804, gave no perceptible differences from the air taken in Manchester.—M. Gay-Lussac determines the constitution of air brought from an elevation of four miles to be the same as that at the earth's surface.

† From the Memoirs of the Literary and Philosophical Society of Manchester, Second Series, Volume I., 1805, pp. 271-287. The table appended to this paper is Dalton's first table of atomic weights.

tion, as exhibited in the following table :—It must be understood that the quantity of gas is to be measured *at the pressure and temperature with which the impregnation is effected.*

Bulk absorbed, the bulk of water being unity. $\frac{1}{1^3} = 1$	Carbonic acid gas, sulphuretted hydrogen, nitrous oxide.*
$\frac{1}{2^3} = \frac{1}{8}$	Olefiant gas, of the Dutch chemists.
$\frac{1}{3^3} = \frac{1}{27}$	Oxygenous gas, nitrous gas,† carburetted hydrogen gas, from stagnant water.
$\frac{1}{4^3} = \frac{1}{64}$	Azotic gas, hydrogenous gas, carbonic oxide.
$\frac{1}{5^3} = \frac{1}{125}$	None discovered.

3. The gas thus absorbed may be recovered from the

---

* According to Mr William Henry's experiments, water does not imbibe quite its bulk of nitrous oxide ; in one or two instances with me it has come very near it : The apparent deviation of this gas, may be owing to the difficulty of ascertaining the exact degree of its impurity.

† About $\frac{9}{10}$ of nitrous gas is usually absorbed ; and $\frac{1}{27}$ is recoverable : This difference is owing to the residuum of oxygen in the water, each measure of which takes $3\frac{1}{2}$ of nitrous gas to saturate it, when in water. Perhaps it may be found that nitrous gas usually contains a small portion of nitrous oxide.

water the same in quantity and quality as it entered, by the means pointed out in the 1st article.

4. If a quantity of water free from air be agitated with a mixture of two or more gases (such as atmospheric air) the water will absorb portions of each gas the same as if they were presented to it separately in their proper density.

Ex. gr. Atmospheric air, consisting of 79 parts azotic gas, and 21 parts oxygenous gas, per cent.

$$\text{Water absorbs } \tfrac{1}{64} \text{ of } \tfrac{79}{100}, \text{ azotic gas} = 1.234$$
$$\underline{\hspace{3cm} \tfrac{1}{27} \text{ of } \tfrac{21}{100}, \text{ oxygen gas} = .778}$$

Sum, per cent.   2.012

5. If water impregnated with any one gas (as hydrogenous) be agitated with another gas *equally* absorbable (as azotic) there will *apparently* be no absorption of the latter gas; just as much gas being found after agitation as was introduced to the water; but upon examination the residuary gas will be found a *mixture* of the two, and the parts of each, in the water, will be exactly proportional to those out of the water.

6. If water impregnated with any one gas be agitated with another gas less or more absorbable; there will *apparently* be an increase or diminution of the latter; but upon examination the residuary gas will be found a *mixture* of the two, and the proportions agreeable to article 4.

7. If a quantity of water in a phial having a ground stopper very accurately adapted, be agitated with any gas, or mixture of gases, till the due share has entered the water; then, if the stopper be secured, the phial may be exposed to any variation of *temperature*, without disturbing the equilibrium: That is, the quantity of gas in the water will remain the same whether it be exposed to heat or cold, if the stopper be air-tight.

N.B. The phial ought not to be near full of water, and the temperature should be between 32° and 212°.

8. If water be impregnated with one gas (as oxygenous), and another gas, having an affinity for the former (as nitrous), be agitated along with it ; the absorption of the latter gas will be greater, by the quantity necessary to saturate the former, than it would have been if the water had been free from gas.*

9. Most liquids free from viscidity, such as acids, alcohol, liquid sulphurets, and saline solutions in water, absorb the same quantity of gases as pure water ; except they have an affinity for the gas, such as the sulphurets for oxygen, &c.

The preceding articles contain the principal facts necessary to establish the theory of absorption : Those that follow are of a subordinate nature, and partly deducible as corollaries to them.

10. Pure distilled water, rain and spring water usually contain nearly their due share of atmospheric air : if not, they quickly acquire that share by agitation in it, and lose any other gas they may be impregnated with. It is remarkable however that water by stagnation, in certain circumstances, loses part or all of its oxygen, notwithstanding its constant exposition to the atmosphere. This I have uniformly found to be the case in my large wooden pneumatic trough, containing about 8 gallons, or $1\frac{1}{3}$ cubic foot of water. Whenever this is replenished with tolerably pure rain water, it contains its share of atmospheric

---

* One part of oxygenous gas requires 3.4 of nitrous gas to saturate it in water.   It is agreeable to this that the rapid mixture of oxygenous and nitrous gas over a broad surface of water, occasions a greater diminution than otherwise.   In fact, the *nitrous* acid is formed this way ; whereas, when water is not present, the *nitric* acid is formed, which requires just half the quantity of nitrous gas, as I have lately ascertained.

air ; but in process of time it becomes deficient of oxygen :
In three months the whole surface has been covered with
a pellicle, and no oxygenous gas whatever was found in
the water.   It was grown offensive, but not extremely so ;
it had not been contaminated with any material portion
of metallic or sulphureous mixtures, or any other article
to which the effect could be ascribed.*   The quantity of
azotic gas is not materially diminished by stagnation, if
at all.—These circumstances, not being duly noticed, have
been the source of great diversity in the results of different philosophers upon the quantity and quality of atmospheric air in water.   By article 4, it appears that
atmospheric air expelled from water ought to have 38 per
cent. oxygen ; whereas by this article air may be expelled
from water that shall contain from 38 to o per cent. of
oxygen.—The disappearance of oxygenous gas in water,
I presume, must be owing to some impurities in the water
which combine with the oxygen.   Pure rain water that
had stood more than a year in an earthenware bottle had
lost none of its oxygen.

11. If water free from air be agitated with a small
portion of atmospheric air (as $\frac{1}{15}$ of its bulk) the residuum
of such air will have proportionally less oxygen than the
original : If we take $\frac{1}{15}$, as above, then the residuum will
have only 17 per cent. oxygen ; agreeably to the principle established in article 4.   This circumstance accounts
for the observations made by Dr Priestley, and Mr
William Henry, that water absorbs oxygen in preference
to azot.

12. If a tall glass vessel containing a small portion of
gas be inverted into a deep trough of water and the gas
thus confined by the glass, and the water be briskly
agitated, it will gradually disappear.

---

* It was drawn from a leaden cistern.

It is a wonder that Dr Priestley, who seems to have been the first to notice this fact, should have made any difficulty of it;—the loss of gas has evidently a mechanical cause; the agitation divides the air into an infinite number of minute bubbles which may be seen pervading the whole water; these are successively driven out from under the margin of the glass into the trough, and so escape.

13. If old stagnant water be in the trough, in the last experiment, and atmospheric air be the subject, the oxygenous gas will very soon be almost wholly extracted and leave a residuum of azotic gas; but if the water be fully impregnated with atmospheric air at the beginning, the residuary gas examined at any time will be pure atmospheric air.

14. If any gas not containing either azotic or oxygenous gas, be agitated over water containing atmospheric air, the residuum will be found to contain both azotic and oxygenous gas.

15. Let a quantity of water contain equal portions of any two or more unequally absorbable gases: For instance, azotic gas, oxygenous gas and carbonic acid gas; then, let the water be boiled or subjected to the air-pump, and it will be found that unequal portions of the gases will be expelled. The azotic will be the greatest part, the oxygenous next, and the carbonic acid will be the least.— For, the previous impregnation being such as is due to atmospheres of the following relative forces nearly;

Azotic - - - 21 inch. of mercury.
Oxygenous - - 9
Carbonic acid - $\frac{1}{3}$

consequently, when those forces are removed, the resiliency of the azotic gas will be the greatest, and that of the carbonic acid the least; the last will even be so small as not to overcome the cohesion of the water without violent agitation.

Remarks on the Authority of the preceding Facts.

In order to give the chain of facts as distinct as possible, I have not hitherto mentioned by whom or in what manner they were ascertained.

The fact mentioned in the first article has been long known ; a doubt, however, remained respecting the quantity of air still left in water after ebullition and the operation of the air-pump. The subsequent articles will, I apprehend, have placed this in a clearer point of view.

In determining the quantity of gases absorbed, I had the result of Mr William Henry's experience on the subject before me, an account of which has been published in the Philosophical Transactions for 1803. By the reciprocal communications since, we have been enabled to bring the results of our Experiments to a near agreement ; as the quantities he has given in his appendix to that paper nearly accord with those I have stated in the second article. In my Experiments with the less absorbable gases, or those of the 2d, 3d, and 4th classes, I used a phial holding 2700 grains of water, having a very accurately ground stopper ; in those with the more absorbable of the first class I used an Eudiometer tube properly graduated, and of aperture so as to be covered with the end of a finger. This was filled with the gas and a small portion expelled by introducing a solid body under water ; the quantity being noticed by the quantity of water that entered on withdrawing the solid body, the finger was applied to the end and the water within agitated ; then removing the finger for a moment under water, an additional quantity of water entered, and the agitation was repeated till no more water would enter, when the quantity and quality of the residuary gas was examined. In fact water could never be made to take its bulk of any gas by this procedure ; but if it took $\frac{9}{10}$, or any other part, and the residuary gas was $\frac{9}{10}$ pure, then it was inferred that water would take its bulk of that gas. The principle was the same in using the phial; only a small quantity of the gas was admitted, and the agitation was longer.

There are two very important facts contained in the second article. The first is, that the quantity of gas absorbed is as the density or pressure.—This was discovered by Mr Wm. Henry, before either he or I had formed any theory on the subject.

The other is that the density of the gas in the water has a special relation to that out of the water, the distance of the particles within being always some multiple of that without :—Thus, in the case of carbonic acid, &c. the distance within and without is the same, or the gas within the water is of the same density as without; in olefiant gas the distance of the particles in the water is twice that without; in oxygenous gas, &c. the distance is just three times as great within as without; and in azotic, &c. it is four times. This fact was the result of my own enquiry. The former of these, I think, decides the effect to be mechanical; and the latter seems to point to the principle on which the equilibrium is adjusted.

The facts noticed in the 4th, 5th and 6th articles, were investigated *à priori* from the mechanical hypothesis, and the notion of the distinct agency of elastic fluids when mixed together. The results were found entirely to agree with both, or as nearly as could be expected from experiments of such nature.

The facts mentioned in the 7th article, are of great importance in a theoretic view; for, if the quantity of gas absorbed depend upon mechanical principles, it cannot be affected by temperature in confined air, as the mechanical effect of the external and internal air are alike increased by heat, and the density not at all affected in those circumstances. I have tried the experiments in a considerable variety of temperature without perceiving any deviation from the principle.   It deserves further attention.

If water be, as pointed out by this essay, a mere receptacle of gases, it cannot affect their affinities : hence what is observed in the 8th article is too obvious to need explanation.—And if we find the absorption of gases to arise not from a chemical but a mechanical cause, it may be expected that all liquids having an equal fluidity with water, will absorb like portions of gas.   In several liquids I have tried no perceptible difference has been found ; but this deserves further investigation.

After what has been observed, it seems unnecessary to add any explanation of the 10th and following articles.

Theory of the Absorption of Gases by Water, &c.

From the facts developed in the preceding articles, the following theory of the absorption of gases by water seems deducible.

1. All gases that enter into water and other liquids by means of pressure, and are wholly disengaged again by the removal of that pressure, are *mechanically* mixed with the liquid, and not *chemically* combined with it.

2. Gases so mixed with water, &c. retain their elasticity or repulsive power amongst their own particles, just the same in the water as out of it, the intervening water having no other influence in this respect than a mere vacuum.

3. Each gas is retained in water by the pressure of gas of its own kind incumbent on its surface abstractedly con-sidered, no other gas with which it may be mixed having any permanent influence in this respect.

4. When water has absorbed its bulk of carbonic acid gas, &c., the gas does not press on the water at all, but presses on the containing vessel just as if no water were

in.—When water has absorbed its proper quantity of oxygenous gas, &c. that is, $\frac{1}{27}$ of its bulk, the exterior gas presses on the surface of the water with $\frac{26}{27}$ of its force, and on the internal gas with $\frac{1}{27}$ of its force, which force presses upon the containing vessel and not on the water. With azotic and hydrogenous gas the proportions are $\frac{63}{64}$ and $\frac{1}{64}$ respectively. When water contains no gas, its surface must support the whole pressure of any gas admitted to it, till the gas has, in part, forced its way into the water.

5. A particle of gas pressing on the surface of water is analogous to a single shot pressing upon the summit of a square pile of them. As the shot distributes its pressure equally amongst all the individuals forming the lowest stratum of the pile, so the particle of gas distributes its pressure equally amongst every successive horizontal stratum of particles of water downwards till it reaches the sphere of influence of another particle of gas. For instance; let any gas press with a given force on the surface of water, and let the distance of the particles of gas from each other be to those of water as 10 to 1; then each particle of gas must divide its force equally amongst 100 particles of water, as follows:—It exerts its immediate force upon 4 particles of water; those 4 press upon 9, the 9 upon 16, and so on according to the order of square numbers, till 100 particles of water have the force distributed amongst them; and in the same stratum each square of 100, having its incumbent particle of gas, the water below this stratum is uniformly pressed by the gas, and consequently has not its equilibrium disturbed by that pressure.

6. When water has absorbed $\frac{1}{27}$ of its bulk of any gas, the stratum of gas on the surface of the water presses with $\frac{26}{27}$ of its force on the water, in the manner pointed out in the last article, and with $\frac{1}{27}$ of its force on the uppermost

stratum of gas in the water: The distance of the two strata of gas must be nearly 27 times the distance of the particles in the incumbent atmosphere and 9 times the distance of the particles in the water. This comparatively great distance of the inner and outer atmosphere arises from the great repulsive power of the latter, on account of its superior density, or its presenting 9 particles of surface to the other 1. When $\frac{1}{64}$ is absorbed the distance of the atmospheres becomes 64 times the distance of two particles in the outer, or 16 times that of the inner.

7. An equilibrium between the outer and inner atmospheres can be established in no other circumstance than that of the distance of the particles of one atmosphere being the same or some multiple of that of the other ; and it is probable the multiple cannot be more than 4. For in this case the distance of the inner and outer atmospheres is such as to make the perpendicular force of each particle of the former on those particles of the latter that are immediately subject to its influence, physically speaking, equal ; and the same may be observed of the small lateral force.

8. The greatest difficulty attending the mechanical hypothesis, arises from different gases observing different laws. Why does water not admit its bulk of every kind of gas alike?—This question I have duly considered, and though I am not yet able to satisfy myself completely, I am nearly persuaded that the circumstance depends upon the weight and number of the ultimate particles of the several gases : those whose particles are lightest and single being least absorbable, and the others more according as they increase in weight and complexity.*   An inquiry into the relative weights of the

---

* Subsequent experience renders this conjecture less probable.

ultimate particles of bodies is a subject, as far as I know,
entirely new : I have lately been prosecuting this enquiry
with remarkable success.   The principle cannot be entered
upon in this paper ; but I shall just subjoin the results, as
far as they appear to be ascertained by my experiments.

Table of the relative weights of the ultimate particles of
gaseous and other bodies.

Hydrogen	1
Azot	4.2
Carbone	4.3
Ammonia	5.2
Oxygen	5.5
Water	6.5
Phosphorus	7.2
Phosphuretted hydrogen	8.2
Nitrous gas	9.3
Ether	9.6
Gaseous oxide of carbone	9.8
Nitrous oxide	13.7
Sulphur	14.4
Nitric acid	15.2
Sulphuretted hydrogen	15.4
Carbonic acid	15.3
Alcohol	15.1
Sulphureous acid	19.9
Sulphuric acid	25.4
Carburetted hydrogen from stag. water	6.3
Olefiant gas	5.3

ON THE CONSTITUTION OF BODIES.
By JOHN DALTON.*

THERE are three distinctions in the kinds of bodies, or three states, which have more especially claimed the attention of philosophical chemists; namely, those which are marked by the terms *elastic fluids, liquids, and solids*. A very familiar instance is exhibited to us in water, of a body, which, in certain circumstances, is capable of assuming all the three states. In steam we recognise a perfectly elastic fluid, in water a perfect liquid, and in ice a complete solid. These observations have tacitly led to the conclusion which seems universally adopted, that all bodies of sensible magnitude, whether liquid or solid, are constituted of a vast number of extremely small particles, or atoms of matter bound together by a force of attraction, which is more or less powerful according to circumstances, and which as it endeavours to prevent their separation, is very properly called in that view, *attraction of cohesion;* but as it collects them from a dispersed state (as from steam into water) it is called, *attraction of aggregation*, or more simply, *affinity*. Whatever names it may go by, they still signify one and the same power. It is not my design to call in question this conclusion, which appears completely satisfactory; but to show that we have hitherto made no use of it, and that the consequence of the neglect, has been a very obscure view of chemical agency, which is daily growing more so in proportion to the new lights attempted to be thrown upon it.

The opinions I more particularly allude to, are those of Berthollet on the Laws of chemical affinity; such as that chemical agency is proportional to the mass, and that in all chemical unions, there exist insensible grada-

* From A New System of Chemical Philosophy, Manchester 1808, pp. 141-143.

tions in the proportions of the constituent principles. The inconsistence of these opinions, both with reason and observation, cannot, I think, fail to strike every one who takes a proper view of the phenomena.

Whether the ultimate particles of a body, such as water, are all alike, that is, of the same figure, weight, &c. is a question of some importance. From what is known, we have no reason to apprehend a diversity in these particulars : if it does exist in water, it must equally exist in the elements constituting water, namely, hydrogen and oxygen. Now it is scarcely possible to conceive how the aggregates of dissimilar particles should be so uniformly the same. If some of the particles of water were heavier than others, if a parcel of the liquid on any occasion were constituted principally of these heavier particles, it must be supposed to affect the specific gravity of the mass, a circumstance not known. Similar observations may be made on other substances. Therefore we may conclude that *the ultimate particles of all homogeneous bodies are perfectly alike in weight, figure, &c.* In other words, every particle of water is like every other particle of water ; every particle of hydrogen is like every other particle of hydrogen, &c.

ON CHEMICAL SYNTHESIS.
By JOHN DALTON.*

WHEN any body exists in the elastic state, its ultimate particles are separated from each other to a much greater distance than in any other state ; each particle occupies the centre of a comparatively large sphere, and supports its dignity by keeping all the rest,

* From A New System of Chemical Philosophy, Manchester 1808, pp. 211-216 and 219-220.

which by their gravity, or otherwise are disposed to encroach upon it, at a respectful distance. When we attempt to conceive the *number* of particles in an atmosphere, it is somewhat like attempting to conceive the number of stars in the universe; we are confounded with the thought. But if we limit the subject, by taking a given volume of any gas, we seem persuaded that, let the divisions be ever so minute, the number of particles must be finite; just as in a given space of the universe, the number of stars and planets cannot be infinite.

Chemical analysis and synthesis go no farther than to the separation of particles one from another, and to their re-union. No new creation or destruction of matter is within the reach of chemical agency. We might as well attempt to introduce a new planet into the solar system, or to annihilate one already in existence, as to create or destroy a particle of hydrogen. All the changes we can produce, consist in separating particles that are in a state of cohesion or combination, and joining those that were previously at a distance.

In all chemical investigations, it has justly been considered an important object to ascertain the relative *weights* of the simples which constitute a compound. But unfortunately the enquiry has terminated here; whereas from the relative weights in the mass, the relative weights of the ultimate particles or atoms of the bodies might have been inferred, from which their number and weight in various other compounds would appear, in order to assist and to guide future investigations, and to correct their results. Now it is one great object of this work, to show the importance and advantage of ascertaining *the relative weights of the ultimate particles, both of simple and compound bodies, the number of simple elementary particles which constitute one compound particle, and the number of less compound particles which enter into the formation of one more compound particle.*

If there are two bodies, A and B, which are disposed to combine, the following is the order in which the combinations may take place, beginning with the most simple : namely,

1 atom of A + 1 atom of B = 1 atom of C, binary.
1 atom of A + 2 atoms of B = 1 atom of D, ternary.
2 atoms of A + 1 atom of B = 1 atom of E, ternary.
1 atom of A + 3 atoms of B = 1 atom of F, quaternary.
3 atoms of A + 1 atom of B = 1 atom of G, quaternary.
&c. &c.

The following general rules may be adopted as guides in all our investigations respecting chemical synthesis.

1st. When only one combination of two bodies can be obtained, it must be presumed to be a *binary* one, unless some cause appear to the contrary.

2d. When two combinations are observed, they must be presumed to be a *binary* and a *ternary*.

3d. When three combinations are obtained, we may expect one to be a *binary*, and the other two *ternary*.

4th. When four combinations are observed, we should expect one *binary*, two *ternary*, and one *quaternary*, &c.

5th. A *binary* compound should always be specifically heavier than the mere mixture of its two ingredients.

6th. A *ternary* compound should be specifically heavier than the mixture of a binary and a simple, which would, if combined, constitute it ; &c.

7th. The above rules and observations equally apply, when two bodies, such as C and D, D and E, &c. are combined.

From the application of these rules, to the chemical facts already well ascertained, we deduce the following conclusions; 1st. That water is a binary compound of hydrogen and oxygen, and the relative weights of the two elementary atoms are as 1 : 7, nearly ; 2d. That ammonia is a binary compound of hydrogen and azote, and the relative weights of the two atoms are as 1 : 5, nearly ; 3d. That nitrous gas is a binary compound of azote and oxygen, the atoms of which weigh 5 and 7 respectively; that nitric acid is a binary or ternary compound according as it is derived, and consists of one atom of azote and two of oxygen, together weighing 19; that nitrous oxide is a compound similar to nitric acid, and consists of one atom of oxygen and two of azote, weighing 17; that nitrous acid is a binary compound of nitric acid and nitrous gas, weighing 31; that oxynitric acid is a binary compound of nitric acid and oxygen, weighing 26; 4th. That carbonic oxide is a binary compound, consisting of one atom of charcoal, and one of oxygen, together weighing nearly 12; that carbonic acid is a ternary compound (but sometimes binary) consisting of one atom of charcoal, and two of oxygen, weighing 19; &c. &c. In all these cases the weights are expressed in atoms of hydrogen, each of which is denoted by unity.

In the sequel, the facts and experiments from which these conclusions are derived, will be detailed; as well as a great variety of others from which are inferred the constitution and weight of the ultimate particles of the principal acids, the alkalis, the earths, the metals, the metallic oxides and sulphurets, the long train of neutral salts, and in short, all the chemical compounds which have hitherto obtained a tolerably good analysis. Several of the conclusions will be supported by original experiments.

From the novelty as well as importance of the ideas suggested in this chapter, it is deemed expedient to give plates, exhibiting the mode of combination in some of the more simple cases. A specimen of these accompanies this first part. The elements or atoms of such bodies as are conceived at present to be simple, are denoted by a small circle, with some distinctive mark; and the combinations consist in the juxta-position of two or more of these; when three or more particles of elastic fluids are combined together in one, it is to be supposed that the

particles of the same kind repel each other, and therefore
take their stations accordingly.

ELEMENTS.

Simple

Binary

Ternary

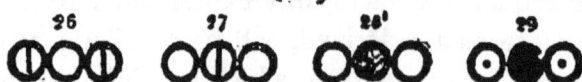

Quaternary

Quinquenary & Sextenary

Septenary

This plate contains the arbitrary marks or signs chosen to represent the several chemical elements or ultimate particles.

Fig.				Fig.				
1	Hydrog. its rel. weight		1	11	Strontites,	-	-	- 46
2	Azote, -	- -	5	12	Barytes,	-	-	- 68
3	Carbone or charcoal,	-	5	13	Iron,	-	-	- 38
4	Oxygen,	- -	7	14	Zinc,	-	-	- 56
5	Phosphorus, -	- -	9	15	Copper,	-	-	- 56
6	Sulphur,	- -	13	16	Lead, -	-	-	- 95
7	Magnesia,	- -	20	17	Silver, -	-	-	- 100
8	Lime, -	- -	23	18	Platina,	-	-	- 100
9	Soda, -	- -	28	19	Gold, -	-	-	- 140
10	Potash,	- -	42	20	Mercury,	-	-	- 167

21. An atom of water or steam, composed of 1 of oxygen and 1 of hydrogen, retained in physical contact by a strong affinity, and supposed to be surrounded by a common atmosphere of heat ; its relative weight = - - - 8

22. An atom of ammonia, composed of 1 of azote and 1 of hydrogen - - - - - - - - - 6

23. An atom of nitrous gas, composed of 1 of azote and 1 of oxygen - - - - - - - - - 12

24. An atom of olefiant gas, composed of 1 of carbone and 1 of hydrogen - - - - - - - - - 6

25. An atom of carbonic oxide composed of 1 of carbone and 1 of oxygen - - - - - - - - - 12

26. An atom of nitrous oxide, 2 azote + 1 oxygen - - - 17

27. An atom of nitric acid, 1 azote + 2 oxygen - - - 19

28. An atom of carbonic acid, 1 carbone + 2 oxygen - - 19

29. An atom of carburetted hydrogen, 1 carbone + 2 hydrogen 7

30. An atom of oxynitric acid, 1 azote + 3 oxygen - - 26

31. An atom of sulphuric acid, 1 sulphur + 3 oxygen - - 34

32. An atom of sulphuretted hydrogen, 1 sulphur + 3 hydrogen - - - - - - - - - 16

33. An atom of alcohol, 3 carbone + 1 hydrogen - - - 16

34. An atom of nitrous acid, 1 nitric acid + 1 nitrous gas - 31

35. An atom of acetous acid, 2 carbone + 2 water - - 26

36. An atom of nitrate of ammonia, 1 nitric acid + 1 ammonia + 1 water - - - - - - - - - 33

37. An atom of sugar, 1 alcohol + 1 carbonic acid - - 35

Enough has been given to show the method ; it will be quite unnecessary to devise characters and combinations of them to exhibit to view in this way all the subjects that come under investigation ; nor is it necessary to insist upon the accuracy of all these compounds, both in number and weight ; the principle will be entered into more particularly hereafter, as far as respects the individual results. It is not to be understood that all those articles marked as simple substances, are necessarily such by the theory ; they are only necessarily of such weights. Soda and potash, such as they are found in combination with acids, are 28 and 42 respectively in weight ; but according to Mr Davy's very important discoveries, they are metallic oxides ; the former then must be considered as composed of an atom of metal, 21, and one of oxygen, 7 ; and the latter, of an atom of metal, 35, and one of oxygen, 7. Or, soda contains 75 per cent. metal and 25 oxygen ; potash, 83.3 metal and 16.7 oxygen. It is particularly remarkable, that according to the above-mentioned gentleman's essay on the Decomposition and Composition of the fixed alkalies, in the Philosophical Transactions (a copy of which essay he has just favoured me with) it appears that "the largest quantity of oxygen indicated by these experiments was for potash 17, and for soda, 26 parts in 100, and the smallest 13 and 19."

ON SUPER-ACID AND SUB-ACID SALTS. By WILLIAM HYDE WOLLASTON, M.D., Sec. R.S.*

Read Jan. 28, 1808.

IN the paper which has just been read to the Society, Dr Thomson has remarked, that oxalic acid unites to strontian as well as to potash in two different proportions, and that the quantity of acid combined with each of these bases in their super-oxalates, is just double of that which is saturated by the same quantity of base in their neutral compounds.†

As I had observed the same law to prevail in various other instances of super-acid and sub-acid salts, I thought

* From the Philosophical Transactions, vol. 98 (for 1808), pp. 96-102.

† [See the extracts from Dr Thomson's paper on Oxalic Acid, on page 41.]

it not unlikely that this law might obtain generally in such compounds, and it was my design to have pursued the subject with the hope of discovering the cause to which so regular a relation might be ascribed.

But since the publication of Mr Dalton's theory of chemical combination, as explained and illustrated by Dr Thomson,* the inquiry which I had designed appears to be superfluous, as all the facts that I had observed are but particular instances of the more general observation of Mr Dalton, that in all cases the simple elements of bodies are disposed to unite atom to atom singly, or, if either is in excess, it exceeds by a ratio to be expressed by some simple multiple of the number of its atoms.

However, since those who are desirous of ascertaining the justness of this observation by experiment, may be deterred by the difficulties that we meet with in attempting to determine with precision the constitution of gaseous bodies, for the explanation of which Mr Dalton's theory was first conceived, and since some persons may imagine that the results of former experiments on such bodies do not accord sufficiently to authorize the adoption of a new hypothesis, it may be worth while to describe a few experiments, each of which may be performed with the utmost facility, and each of which affords the most direct proof of the proportional redundance or deficiency of acid in the several salts employed.

Sub-carbonate of Potash.

Exp. 1. Sub-carbonate of potash recently prepared, is one instance of an alkali having one-half the quantity of acid necessary for its saturation, as may thus be satisfactorily proved.

Let two grains of fully saturated and well crystallized carbonate of potash be wrapped in a piece of thin paper,

---

* Thomson's Chemistry, 3d edition, vol. iii., p. 425.

and passed up into an inverted tube filled with mercury, and let the gas be extricated from it by a sufficient quantity of muriatic acid, so that the space it occupies may be marked upon the tube.

Next, let four grains of the same carbonate be exposed for a short time to a red heat ; and it will be found to have parted with exactly half its gas ; for the gas extricated from it in the same apparatus will be found to occupy exactly the same space, as the quantity before obtained from two grains of fully saturated carbonate.

Sub-carbonate of Soda.

Exp. 2. A similar experiment may be made with a saturated carbonate of soda, and with the same result ; for this also becomes a true semi-carbonate by being exposed for a short time to a red heat.

Super-sulphate of Potash.

By an experiment equally simple, super-sulphate of potash may be shown to contain exactly twice as much acid as is necessary for the mere saturation of the alkali present.

Exp. 3. Let twenty grains of carbonate of potash (which would be more than neutralized by ten grains of sulphuric acid) be mixed with about twenty-five grains of that acid in a covered crucible of platina, or in a glass tube three quarters of an inch diameter, and five or six inches long.

By heating this mixture till it ceases to boil, and begins to appear slightly red hot, a part of the redundant acid will be expelled, and there will remain a determinate quantity forming super-sulphate of potash, which when dissolved in water will be very nearly neutralized by an addition of twenty grains more of the same carbonate of potash ; but it is generally found very slightly acid, in consequence of the small quantity of sulphuric acid which remains in the vessel in a gaseous state at a red heat.

In the preceding experiments, the acids are made to assume a determinate proportion to their base, by heat which cannot destroy them. In those which follow, the proportion which a destructible acid shall assume cannot be regulated by the same means; but the constitution of its compounds previously formed, may nevertheless be proved with equal facility.

Super-oxalate of Potash.

Exp. 4. The common super-oxalate of potash is a salt that contains alkali sufficient to saturate exactly half of the acid present Hence, if two equal quantities of salt of sorrel be taken, and if one of them be exposed to a red heat, the alkali which remains will be found exactly to saturate the redundant acid of the other portion.

In addition to the preceding compounds, selected as distinct examples of binacid salts, I have observed one remarkable instance of a more extended and general prevalence of the law under consideration; for when the circumstances are such as to admit the union of a further quantity of oxalic acid with potash, I found a proportion, though different, yet analogous to the former, regularly to occur.

§ Quadroxalate of Potash.

In attempting to decompose the preceding super-oxalate by means of acids, it appeared that nitric or muriatic acids, are capable of taking only half the alkali, and that the salt which crystallizes after solution in either of these acids, has accordingly exactly four times as much acid as would saturate the alkali that remains.   -

Exp. 5. For the purpose of proving that the constitution of this compound has been rightly ascertained, the salt thus formed should be purified by a second crystallization in distilled water; after which the alkali of thirty grains must be obtained by exposure to a red heat, in order to neutralize the redundant acid contained in ten grains of the same salt. The quantity of unburned salt contains alkali for one part out of four of the acid present, and it requires the alkali of three equal quantities of the same salt to saturate the three remaining parts of acid.

The limit to the decomposition of super-oxalate of potash by the above acids, is analogous to that which occurs when sulphate of potash is decomposed by nitric acid; for in this case also, no quantity of that acid can take more than half the potash, and the remaining salt is converted into a definite super-sulphate, similar to that obtained by heat in the third experiment.

It is not improbable that many other changes in chemistry, supposed to be influenced by a general redundance of some one ingredient, may in fact be limited by a new order of affinities taking place at some definite proportion to be expressed by a simple multiple. And though the strong power of crystallizing in oxalic acid, renders the modifications of which its combinations are susceptible more distinct than those of other acids, it seems probable that a similar play of affinities will arise in solution, when other acids exceed their base in the same proportion.

In order to determine whether oxalic acid is capable of uniting to potash in a proportion intermediate between the double and quadruple quantity of acid, I neutralized forty-eight grains of carbonate of potash with thirty grains of oxalic acid, and added sixty grains more of acid, so that I had two parts of potash of twenty-four grains each, and six *equivalent* quantities of oxalic acid of fifteen grains each, in solution, ready to crystallize together, if disposed to unite, in the proportion of three to one; but the first

portion of salt that crystallized, was the common binoxalate, or salt of sorrel, and a portion selected from the
after crystals (which differed very discernibly in their
form) was found to contain the quadruple proportion of
acid. Hence it is to be presumed, that if these salts
could have been perfectly separated, it would have been
found, that the two quantities of potash were equally
divided, and combined in one instance with two, and in
the other with the remaining four out of the six *equivalent*
quantities of acid taken.

To account for this want of disposition to unite in the
proportion of three to one by Mr Dalton's theory, I
apprehend he might consider the neutral salt as consisting of

      2 particles potash with 1 acid,
The binoxalate
   as 1 and 1, or 2   with 2,
The quadroxalate
   as 1 and 2, or 2   with 4,
in which cases the ratios which I have observed of the
acids to each other in these salts would respectively
obtain.

But an explanation, which admits the supposition of a
double share of potash in the neutral salt, is not altogether
satisfactory ; and I am further inclined to think, that when
our views are sufficiently extended, to enable us to reason
with precision concerning the proportions of elementary
atoms, we shall find the arithmetical relation alone will
not be sufficient to explain their mutual action, and that
we shall be obliged to acquire a geometrical conception
of their relative arrangement in all the three dimensions
of solid extension.

For instance, if we suppose the limit to the approach
of particles to be the same in all directions, and hence
their virtual extent to be spherical (which is the most

simple hypothesis); in this case, when different sorts combine singly there is but one mode of union. If they unite in the proportion of two to one, the two particles will naturally arrange themselves at opposite poles of that to which they unite. If there be three, they might be arranged with regularity, at the angles of an equilateral triangle in a great circle surrounding the single spherule; but in this arrangement, for want of similar matter at the poles of this circle, the equilibrium would be unstable, and would be liable to be deranged by the slightest force of adjacent combinations; but when the number of one set of particles exceeds in the proportion of four to one, then, on the contrary, a stable equilibrium may again take place, if the four particles are situated at the angles of the four equilateral triangles composing a regular tetra-hedron.

But as this geometrical arrangement of the primary elements of matter is altogether conjectural, and must rely for its confirmation or rejection upon future inquiry, I am desirous that it should not be confounded with the results of the facts and observations related above, which are sufficiently distinct and satisfactory with respect to the existence of the law of simple multiples. It is perhaps too much to hope, that the geometrical arrangement of primary particles will ever be perfectly known; since even admitting that a very small number of these atoms com-bining together would have a tendency to arrange them-selves in the manner I have imagined; yet, until it is ascertained how small a proportion the primary particles themselves bear to the interval between them, it may be supposed that surrounding combinations, although them-selves analogous, might disturb that arrangement, and in that case, the effect of such interference must also be taken into the account, before any theory of chemical combination can be rendered complete.

EXTRACTS FROM A PAPER ON OXALIC ACID. By THOMAS THOMSON, M.D., F.R.S.Ed.*

Read Jan. 14, 1808.

Pp. 69-70.—Oxalate of potash readily crystallizes in flat rhomboids, commonly terminated by dihedral summits. The lateral edges of the prism are usually bevelled. The taste of this salt is cooling and bitter. At the temperature of 60° it dissolves in thrice its weight of water. When dried on the sand bath, and afterwards exposed in a damp place, it absorbs a little moisture from the atmosphere.

This salt combines with an excess of acid, and forms a superoxalate, long known by the name of *salt of sorrel.* It is very sparingly soluble in water, though more so than tartar. It occurs in commerce in beautiful 4-sided prisms attached to each other. The acid contained in this salt is very nearly double of what is contained in oxalate of potash. Suppose 100 parts of potash; if the weight of acid necessary to convert this quantity into oxalate be x, then $2x$ will convert it into superoxalate.

P. 74, . . . It appears that there are two oxalates of strontian, the first obtained by saturating oxalic acid with strontian water, the second by mixing together oxalate of ammonia and muriate of strontian. It is remarkable that the first contains just double the proportion of base contained in the second.

---

* Philosophical Transactions, vol. 98 (for 1808), p. 63.

EXTRACTS FROM THOMSON'S SYSTEM OF CHEMISTRY.*

This difference between the density of the gases, while their elasticity is the same, must be owing to one of two causes : Either the *repulsive force*, or the *density* of the atoms, differs in different gases. The first supposition is by no means probable, supposing the size and density of the particles of different gases the same, and indeed would but ill agree with the analogy of nature ; but the second is very likely to be the true cause. And if we suppose the size and density of the atoms of different gases to differ, this in reality includes the first cause likewise ; for every variation in size and density must necessarily occasion a corresponding variation in the repulsive force, even supposing that force abstractedly considered to be the same in all.

We have no direct means of ascertaining the density of the atoms of bodies ; but Mr Dalton, to whose uncommon ingenuity and sagacity the philosophic world is no stranger, has lately contrived an hypothesis which, if it prove correct, will furnish us with a very simple method of ascertaining that density with great precision. Though the author has not yet thought fit to publish his hypothesis, yet as the notions of which it consists are original and extremely interesting, and as they are intimately connected with some of the most intricate parts of the doctrine of affinity, I have ventured, with Mr

---

* A System of Chemistry. By Thomas Thomson, M.D., F.R.S.E., 3d edition, vol. iii., Edinburgh 1807, pp. 424-429 and 451-452. These extracts form a part of the first public exposition of Dalton's views.

Dalton's permission, to enrich this work with a short sketch of it.*

The hypothesis upon which the whole of Mr Dalton's notions respecting chemical elements is founded, is this: When two elements unite to form a third substance, it is to be presumed that *one* atom of one joins to *one* atom of the other, unless when some reason can be assigned for supposing the contrary. Thus oxygen and hydrogen unite together and form water. We are to presume that an atom of water is formed by the combination of *one* atom of oxygen with *one* atom of hydrogen. In like manner *one* atom of ammonia is formed by the combination of *one* atom of azote with *one* atom of hydrogen. If we represent an atom of oxygen, hydrogen, and azote, by the following symbols,

Oxygen - - - - - - ◯

Hydrogen - - - - - ⊙

Azote - - - - - - ⦶

Then an atom of water and of ammonia will be represented respectively by the following symbols:

Water - - - - - ◯⊙

Ammonia - - - - ⊙⦶

But if this hypothesis be allowed, it furnishes us with a ready method of ascertaining the relative density of those atoms that enter into such combinations; for it has been proved by analysis, that water is composed of

* In justice to Mr Dalton, I must warn the reader not to decide upon the notions of that philosopher from the sketch which I have given, derived from a few minutes' conversation, and from a short written memorandum. The mistakes, if any occur, are to be laid to my account, and not to his; as it is extremely probable that I may have misconceived his meaning in some points.

$85\frac{2}{3}$ of oxygen and $14\frac{1}{3}$ of hydrogen. An atom of water of course is composed of $85\frac{2}{3}$ parts by weight of oxygen and $14\frac{1}{3}$ parts of hydrogen. Now, if it consist of one atom of oxygen united to one atom of hydrogen, it follows, that the weight of one atom of hydrogen is to that of one atom of oxygen as $14\frac{1}{3}$ to $85\frac{2}{3}$, or as 1 to 6 very nearly. In like manner an atom of ammonia has been shown to consist of 80 parts of azote and 20 of hydrogen. · Hence an atom of hydrogen is to an atom of azote as 20 to 80, or as 1 to 4. Thus we have obtained the following relative densities of these three elementary bodies.

Hydrogen - - - - - - 1
Azote - - - - - - 4
Oxygen - - - - - - 6

We have it in our power to try how far this hypothesis is consonant to experiment, by examining the combination of azote and oxygen, on the supposition that these bodies unite, atom to atom, and that the respective densities of the atoms are as in the preceding table. But azote and oxygen unite in various proportions, forming nitrous oxide, nitrous gas, and nitric acid, besides some other compounds which need not be enumerated. The preceding hypothesis will not apply to all these compounds; Mr Dalton, therefore, extends it farther. Whenever more than one compound is formed by the combination of two elements, then the next simple combination must, he supposes, arise from the union of *one* atom of the one with *two* atoms of the other. If we suppose *nitrous gas*, for example, to be composed of *one* atom of azote, and *one* of oxygen, we shall have two new compounds, by uniting an atom of nitrous gas to an atom of azote, and to an atom of oxygen, respectively. If we suppose farther, that nitrous oxide is composed of an atom of nitrous gas and an atom of azote, while

nitric acid consists of nitrous gas and oxygen, united
atom to atom; then the following will be the symbols
and constituents of these three bodies :

Nitrous gas - - - - ⊙◑

Nitrous oxide - - - - ◐⊙◑

Nitric acid - - - - ⊙◑◯

The first gas consists only of two atoms, or is a binary
compound, but the two others consist of three atoms,
or are ternary compounds; nitrous oxide contains two
atoms of azote united to one of oxygen, while nitric
acid consists of two atoms of oxygen united to one of
azote.

When the atoms of two elastic fluids join together to
form *one* atom of a new elastic fluid, the density of this
new compound is always greater than the mean. Thus
the density of nitrous gas, by calculation, ought only to
be 1.045; but its real density is 1.094. Now as both
nitrous oxide and nitric acid are specifically heavier than
nitrous gas, though the one contains more of the lighter
ingredient, and the other more of the heavier ingredient
than that compound does, it is reasonable to conclude,
that they are combinations of nitrous gas with azote and
oxygen respectively, and that this is the reason of the
increased specific gravity of each; whereas were not this
the case, nitrous oxide ought to be specifically lighter than
nitrous gas. Supposing, then, the constituents of these
gases to be as represented in the preceding table, let us
see how far this analysis will correspond with the densities of their elements, as above deduced from the compo-
sitions of water and ammonia.

Nitrous gas is composed of 1.00 azote and 1.36
oxygen, or of 4 azote and 5.4 oxygen.

Nitrous oxide, of 2 azote and 1.174 oxygen, or of
4 + 4 azote and 4.7 oxygen.

Nitric acid, of 1 azote and 2.36 oxygen, or of 4 azote and 4.7 + 4.7 oxygen.

These three give us the following as the relative densities of azote and oxygen :

Azote.		Oxygen.
4	:	5.4
	:	4.7
	:	4.7

The mean of the whole is nearly 4 : 5 ; but from preceding analyses of water and ammonia, we obtained their densities 4 : 6. Though these results do not correspond exactly, yet the difference is certainly not very great, and indeed as little as can reasonably be expected, even supposing the hypothesis is well founded, if we consider the extreme difficulty of attaining accuracy in the analysis of gaseous compounds. If ammonia were supposed a compound of 83 azote and 17 hydrogen, instead of 80 azote and 20 hydrogen, in that case the density of azote would be five instead of four, and the different sets of experiments would coincide very nearly. Now it is needless to observe how easy it is, in analysing gaseous compounds, to commit an error of 3 *per cent.* which is all that would be necessary to make the different numbers tally.

On the supposition that the hypothesis of Mr Dalton is well founded, the following table exhibits the density of the atoms of the simple gases, and of those which are composed of elastic fluids, together with the symbols of the composition of these compound atoms :

	Hydrogen	-	-	-	-	1
	Azote	-	-	-	-	5
	Oxygen	-	-	-	-	6
	Muriatic acid	-		-		9

⊙◯	Water - - - -	7
⊙◖	Ammonia - · -	6
◯◖	Nitrous gas - -	11
◖◯◖	Nitrous oxide - - -	16
◯◖◯	Nitric acid - · -	17
⊖◯⊖	Oxymuriatic acid - -	24
◯◖◯ Hyperoxymuriatic acid - 27		

* * *

Oxygen and nitrous gas, when brought into contact, immediately unite and form a yellow-coloured vapour. From the experiments of Dalton, it appears that these gases are capable of uniting in two different proportions. One hundred measures of common air, when added to 36 measures of nitrous gas in a narrow tube over water, leave a residue of 79 inches; and one hundred measures of common air, admitted to 72 measures of nitrous gas in a wide glass vessel over water, leave likewise a residue of 79 measures.* According to these experiments, 21 cubic inches of oxygen gas is capable of uniting with 36 and with 72 cubic inches of nitrous gas; or 100 inches of oxygen unite with 171.4 and with 342.8 inches of nitrous gas. If we apply Mr Dalton's hypothesis, stated in a former Section, to these combinations, we shall have the first composed of one atom of oxygen united to one atom of nitrous gas; the second, of one atom of oxygen united to two atoms of nitrous gas. The first appears to be the substance usually distinguished by the name of nitric acid; the second is nitrous vapour, or nitric acid saturated with nitrous gas. The following will be the

* Phil. Mag. xxiii. 351.

symbols denoting the composition of an atom of each, and the density of that atom, obtained by adding together the numbers denoting the density of each of the constituent atoms.

Density.

◯◖◗◯ Nitric acid -   -   - 17

◯◖◗◯◖◗◯ Nitrous vapour -   - 28

The first is a triple compound, and can only be resolved into nitrous gas and oxygen, or into azote and oxygen; but the second is a quintuple compound, and may be resolved into nitrous oxide and oxygen, nitrous gas and oxygen, nitric acid and nitrous gas, oxygen and azote.

www.ingramcontent.com/pod-product-compliance
Lightning Source LLC
Chambersburg PA
CBHW022026190326
41519CB00010B/1611